THIS BOOK

BELONGS

TO

DEDICATION

This Call Log is dedicated to all the personal assistants out there who want to track their phone messages and document their findings in the process.

You are my inspiration for producing books and I'm honored to be a part of keeping all of your Call Log notes, and records organized.

This journal notebook will help you record your details about tracking your messages and voice mails.

Thoughtfully put together with these sections to record: Date & TIme, Whom The Call & Message Is For, Delivered Tick Box, Caller Name, Position, Company, Phone Number, Email, Notes For Message & more!

HOW TO USE THIS BOOK

The purpose of this book is to keep all of your Call Log & Message notes all in one place. It will help keep you organized.

This Call Log will allow you to accurately document every detail about all of your Phone Messages. It's a great way to chart your course through organizing all your Calls.

Here are examples of the prompts for you to fill in and write about your experience in this book:

1. Date & Time

2. Whom The Call & Message Is For

3. Delivered Tick Box

4. Caller Name, Position, Company, Phone Number, Cell Number, Email

5. Notes For The Message

6. Type Of Call: Urgent, Call, Returning Call, Will Call Back (Followup), Came By, Wants To See You

Delivered? ☐	For	Date	Time

Caller	Position	Company

Phone	Cell	Email

Message	Type of call
	☐ Urgent
	☐ Call
	☐ Returning Call
	☐ Please Call
	☐ Will Call Again
	☐ Came to see you
	☐ Wants to see you

Delivered? ☐	For	Date	Time

Caller	Position	Company

Phone	Cell	Email

Message	Type of call
	☐ Urgent
	☐ Call
	☐ Returning Call
	☐ Please Call
	☐ Will Call Again
	☐ Came to see you
	☐ Wants to see you

Delivered? ☐	For	Date	Time

Caller	Position	Company

Phone	Cell	Email

Message	Type of call
	☐ Urgent
	☐ Call
	☐ Returning Call
	☐ Please Call
	☐ Will Call Again
	☐ Came to see you
	☐ Wants to see you

Delivered? ☐	For	Date	Time

Caller	Position	Company

Phone	Cell	Email

Message	Type of call
	☐ Urgent
	☐ Call
	☐ Returning Call
	☐ Please Call
	☐ Will Call Again
	☐ Came to see you
	☐ Wants to see you

Delivered? ☐	For	Date	Time

Caller	Position	Company

Phone	Cell	Email

Message	Type of call
	☐ Urgent
	☐ Call
	☐ Returning Call
	☐ Please Call
	☐ Will Call Again
	☐ Came to see you
	☐ Wants to see you

Delivered? ☐	For	Date	Time

Caller	Position	Company

Phone	Cell	Email

Message	Type of call
	☐ Urgent
	☐ Call
	☐ Returning Call
	☐ Please Call
	☐ Will Call Again
	☐ Came to see you
	☐ Wants to see you

Delivered? ☐	For	Date	Time

Caller	Position	Company

Phone	Cell	Email

Message	Type of call
	☐ Urgent
	☐ Call
	☐ Returning Call
	☐ Please Call
	☐ Will Call Again
	☐ Came to see you
	☐ Wants to see you

Delivered? ☐	For	Date	Time

Caller	Position	Company

Phone	Cell	Email

Message	Type of call
	☐ Urgent
	☐ Call
	☐ Returning Call
	☐ Please Call
	☐ Will Call Again
	☐ Came to see you
	☐ Wants to see you

Delivered? ☐	For	Date	Time

Caller	Position	Company

Phone	Cell	Email

Message	Type of call
	☐ Urgent
	☐ Call
	☐ Returning Call
	☐ Please Call
	☐ Will Call Again
	☐ Came to see you
	☐ Wants to see you

Delivered? ☐	For	Date	Time

Caller	Position	Company

Phone	Cell	Email

Message	Type of call
	☐ Urgent
	☐ Call
	☐ Returning Call
	☐ Please Call
	☐ Will Call Again
	☐ Came to see you
	☐ Wants to see you

Delivered? ☐	For	Date	Time

Caller	Position	Company

Phone	Cell	Email

Message	Type of call
	☐ Urgent
	☐ Call
	☐ Returning Call
	☐ Please Call
	☐ Will Call Again
	☐ Came to see you
	☐ Wants to see you

Delivered? ☐	For	Date	Time

Caller	Position	Company

Phone	Cell	Email

Message	Type of call
	☐ Urgent
	☐ Call
	☐ Returning Call
	☐ Please Call
	☐ Will Call Again
	☐ Came to see you
	☐ Wants to see you

Delivered? ☐	For	Date	Time

Caller	Position	Company

Phone	Cell	Email

Message	Type of call
	☐ Urgent ☐ Call ☐ Returning Call ☐ Please Call ☐ Will Call Again ☐ Came to see you ☐ Wants to see you

Delivered? ☐	For	Date	Time

Caller	Position	Company

Phone	Cell	Email

Message	Type of call
	☐ Urgent ☐ Call ☐ Returning Call ☐ Please Call ☐ Will Call Again ☐ Came to see you ☐ Wants to see you

Delivered? ☐	For	Date	Time

Caller	Position	Company

Phone	Cell	Email

Message	Type of call
	☐ Urgent ☐ Call ☐ Returning Call ☐ Please Call ☐ Will Call Again ☐ Came to see you ☐ Wants to see you

Delivered? ☐	For	Date	Time

Caller	Position	Company

Phone	Cell	Email

Message	Type of call
	☐ Urgent
	☐ Call
	☐ Returning Call
	☐ Please Call
	☐ Will Call Again
	☐ Came to see you
	☐ Wants to see you

Delivered? ☐	For	Date	Time

Caller	Position	Company

Phone	Cell	Email

Message	Type of call
	☐ Urgent
	☐ Call
	☐ Returning Call
	☐ Please Call
	☐ Will Call Again
	☐ Came to see you
	☐ Wants to see you

Delivered? ☐	For	Date	Time

Caller	Position	Company

Phone	Cell	Email

Message	Type of call
	☐ Urgent
	☐ Call
	☐ Returning Call
	☐ Please Call
	☐ Will Call Again
	☐ Came to see you
	☐ Wants to see you

Delivered? ☐	For	Date	Time

Caller	Position	Company

Phone	Cell	Email

Message	Type of call
	☐ Urgent
	☐ Call
	☐ Returning Call
	☐ Please Call
	☐ Will Call Again
	☐ Came to see you
	☐ Wants to see you

Delivered? ☐	For	Date	Time

Caller	Position	Company

Phone	Cell	Email

Message	Type of call
	☐ Urgent
	☐ Call
	☐ Returning Call
	☐ Please Call
	☐ Will Call Again
	☐ Came to see you
	☐ Wants to see you

Delivered? ☐	For	Date	Time

Caller	Position	Company

Phone	Cell	Email

Message	Type of call
	☐ Urgent
	☐ Call
	☐ Returning Call
	☐ Please Call
	☐ Will Call Again
	☐ Came to see you
	☐ Wants to see you

Delivered? ☐	For	Date	Time

Caller	Position	Company

Phone	Cell	Email

Message	Type of call
	☐ Urgent
	☐ Call
	☐ Returning Call
	☐ Please Call
	☐ Will Call Again
	☐ Came to see you
	☐ Wants to see you

Delivered? ☐	For	Date	Time

Caller	Position	Company

Phone	Cell	Email

Message	Type of call
	☐ Urgent
	☐ Call
	☐ Returning Call
	☐ Please Call
	☐ Will Call Again
	☐ Came to see you
	☐ Wants to see you

Delivered? ☐	For	Date	Time

Caller	Position	Company

Phone	Cell	Email

Message	Type of call
	☐ Urgent
	☐ Call
	☐ Returning Call
	☐ Please Call
	☐ Will Call Again
	☐ Came to see you
	☐ Wants to see you

Delivered? ☐	For	Date	Time

Caller	Position	Company

Phone	Cell	Email

Message	Type of call
	☐ Urgent
	☐ Call
	☐ Returning Call
	☐ Please Call
	☐ Will Call Again
	☐ Came to see you
	☐ Wants to see you

Delivered? ☐	For	Date	Time

Caller	Position	Company

Phone	Cell	Email

Message	Type of call
	☐ Urgent
	☐ Call
	☐ Returning Call
	☐ Please Call
	☐ Will Call Again
	☐ Came to see you
	☐ Wants to see you

Delivered? ☐	For	Date	Time

Caller	Position	Company

Phone	Cell	Email

Message	Type of call
	☐ Urgent
	☐ Call
	☐ Returning Call
	☐ Please Call
	☐ Will Call Again
	☐ Came to see you
	☐ Wants to see you

Delivered? ☐	For	Date	Time

Caller	Position	Company

Phone	Cell	Email

Message	Type of call
	☐ Urgent
	☐ Call
	☐ Returning Call
	☐ Please Call
	☐ Will Call Again
	☐ Came to see you
	☐ Wants to see you

Delivered? ☐	For	Date	Time

Caller	Position	Company

Phone	Cell	Email

Message	Type of call
	☐ Urgent
	☐ Call
	☐ Returning Call
	☐ Please Call
	☐ Will Call Again
	☐ Came to see you
	☐ Wants to see you

Delivered? ☐	For	Date	Time

Caller	Position	Company

Phone	Cell	Email

Message	Type of call
	☐ Urgent
	☐ Call
	☐ Returning Call
	☐ Please Call
	☐ Will Call Again
	☐ Came to see you
	☐ Wants to see you

Delivered? ☐	For	Date	Time

Caller	Position	Company

Phone	Cell	Email

Message	Type of call
	☐ Urgent
	☐ Call
	☐ Returning Call
	☐ Please Call
	☐ Will Call Again
	☐ Came to see you
	☐ Wants to see you

Delivered? ☐	For	Date	Time

Caller	Position	Company

Phone	Cell	Email

Message	Type of call
	☐ Urgent
	☐ Call
	☐ Returning Call
	☐ Please Call
	☐ Will Call Again
	☐ Came to see you
	☐ Wants to see you

Delivered? ☐	For	Date	Time

Caller	Position	Company

Phone	Cell	Email

Message	Type of call
	☐ Urgent
	☐ Call
	☐ Returning Call
	☐ Please Call
	☐ Will Call Again
	☐ Came to see you
	☐ Wants to see you

Delivered? ☐	For	Date	Time

Caller	Position	Company

Phone	Cell	Email

Message	Type of call
	☐ Urgent
	☐ Call
	☐ Returning Call
	☐ Please Call
	☐ Will Call Again
	☐ Came to see you
	☐ Wants to see you

Delivered? ☐	For	Date	Time

Caller	Position	Company

Phone	Cell	Email

Message	Type of call
	☐ Urgent
	☐ Call
	☐ Returning Call
	☐ Please Call
	☐ Will Call Again
	☐ Came to see you
	☐ Wants to see you

Delivered? ☐	For	Date	Time

Caller	Position	Company

Phone	Cell	Email

Message	Type of call
	☐ Urgent
	☐ Call
	☐ Returning Call
	☐ Please Call
	☐ Will Call Again
	☐ Came to see you
	☐ Wants to see you

Delivered? ☐	For	Date	Time

Caller	Position	Company

Phone	Cell	Email

Message	Type of call
	☐ Urgent
	☐ Call
	☐ Returning Call
	☐ Please Call
	☐ Will Call Again
	☐ Came to see you
	☐ Wants to see you

Delivered? ☐	For	Date	Time

Caller	Position	Company

Phone	Cell	Email

Message	Type of call
	☐ Urgent
	☐ Call
	☐ Returning Call
	☐ Please Call
	☐ Will Call Again
	☐ Came to see you
	☐ Wants to see you

Delivered? ☐	For	Date	Time

Caller	Position	Company

Phone	Cell	Email

Message	Type of call
	☐ Urgent
	☐ Call
	☐ Returning Call
	☐ Please Call
	☐ Will Call Again
	☐ Came to see you
	☐ Wants to see you

Delivered? ☐	For	Date	Time

Caller	Position	Company

Phone	Cell	Email

Message	Type of call
	☐ Urgent
	☐ Call
	☐ Returning Call
	☐ Please Call
	☐ Will Call Again
	☐ Came to see you
	☐ Wants to see you

Delivered? ☐	For	Date	Time

Caller	Position	Company

Phone	Cell	Email

Message	Type of call
	☐ Urgent
	☐ Call
	☐ Returning Call
	☐ Please Call
	☐ Will Call Again
	☐ Came to see you
	☐ Wants to see you

Delivered? ☐	For	Date	Time

Caller	Position	Company

Phone	Cell	Email

Message	Type of call
	☐ Urgent
	☐ Call
	☐ Returning Call
	☐ Please Call
	☐ Will Call Again
	☐ Came to see you
	☐ Wants to see you

Delivered? ☐	For	Date	Time

Caller	Position	Company

Phone	Cell	Email

Message	Type of call
	☐ Urgent
	☐ Call
	☐ Returning Call
	☐ Please Call
	☐ Will Call Again
	☐ Came to see you
	☐ Wants to see you

Delivered? ☐	For	Date	Time

Caller	Position	Company

Phone	Cell	Email

Message	Type of call
	☐ Urgent
	☐ Call
	☐ Returning Call
	☐ Please Call
	☐ Will Call Again
	☐ Came to see you
	☐ Wants to see you

Delivered? ☐	For	Date	Time

Caller	Position	Company

Phone	Cell	Email

Message	Type of call
	☐ Urgent
	☐ Call
	☐ Returning Call
	☐ Please Call
	☐ Will Call Again
	☐ Came to see you
	☐ Wants to see you

Delivered? ☐	For	Date	Time

Caller	Position	Company

Phone	Cell	Email

Message	Type of call
	☐ Urgent
	☐ Call
	☐ Returning Call
	☐ Please Call
	☐ Will Call Again
	☐ Came to see you
	☐ Wants to see you

Delivered? ☐	For	Date	Time

Caller	Position	Company

Phone	Cell	Email

Message	Type of call
	☐ Urgent
	☐ Call
	☐ Returning Call
	☐ Please Call
	☐ Will Call Again
	☐ Came to see you
	☐ Wants to see you

Delivered? ☐	For	Date	Time

Caller	Position	Company

Phone	Cell	Email

Message	Type of call
	☐ Urgent
	☐ Call
	☐ Returning Call
	☐ Please Call
	☐ Will Call Again
	☐ Came to see you
	☐ Wants to see you

Delivered? ☐	For	Date	Time

Caller	Position	Company

Phone	Cell	Email

Message	Type of call
	☐ Urgent
	☐ Call
	☐ Returning Call
	☐ Please Call
	☐ Will Call Again
	☐ Came to see you
	☐ Wants to see you

Delivered? ☐	For	Date	Time

Caller	Position	Company

Phone	Cell	Email

Message	Type of call
	☐ Urgent
	☐ Call
	☐ Returning Call
	☐ Please Call
	☐ Will Call Again
	☐ Came to see you
	☐ Wants to see you

Delivered? ☐	For	Date	Time

Caller	Position	Company

Phone	Cell	Email

Message	Type of call
	☐ Urgent
	☐ Call
	☐ Returning Call
	☐ Please Call
	☐ Will Call Again
	☐ Came to see you
	☐ Wants to see you

Delivered? ☐	For	Date	Time

Caller	Position	Company

Phone	Cell	Email

Message	Type of call
	☐ Urgent
	☐ Call
	☐ Returning Call
	☐ Please Call
	☐ Will Call Again
	☐ Came to see you
	☐ Wants to see you

Delivered? ☐	For	Date	Time

Caller	Position	Company

Phone	Cell	Email

Message	Type of call
	☐ Urgent
	☐ Call
	☐ Returning Call
	☐ Please Call
	☐ Will Call Again
	☐ Came to see you
	☐ Wants to see you

Delivered? ☐	For	Date	Time

Caller	Position	Company

Phone	Cell	Email

Message	Type of call
	☐ Urgent
	☐ Call
	☐ Returning Call
	☐ Please Call
	☐ Will Call Again
	☐ Came to see you
	☐ Wants to see you

Delivered? ☐	For	Date	Time

Caller	Position	Company

Phone	Cell	Email

Message	Type of call
	☐ Urgent ☐ Call ☐ Returning Call ☐ Please Call ☐ Will Call Again ☐ Came to see you ☐ Wants to see you

Delivered? ☐	For	Date	Time

Caller	Position	Company

Phone	Cell	Email

Message	Type of call
	☐ Urgent ☐ Call ☐ Returning Call ☐ Please Call ☐ Will Call Again ☐ Came to see you ☐ Wants to see you

Delivered? ☐	For	Date	Time

Caller	Position	Company

Phone	Cell	Email

Message	Type of call
	☐ Urgent ☐ Call ☐ Returning Call ☐ Please Call ☐ Will Call Again ☐ Came to see you ☐ Wants to see you

Delivered? ☐	For	Date	Time

Caller	Position	Company

Phone	Cell	Email

Message	Type of call
	☐ Urgent
	☐ Call
	☐ Returning Call
	☐ Please Call
	☐ Will Call Again
	☐ Came to see you
	☐ Wants to see you

Delivered? ☐	For	Date	Time

Caller	Position	Company

Phone	Cell	Email

Message	Type of call
	☐ Urgent
	☐ Call
	☐ Returning Call
	☐ Please Call
	☐ Will Call Again
	☐ Came to see you
	☐ Wants to see you

Delivered? ☐	For	Date	Time

Caller	Position	Company

Phone	Cell	Email

Message	Type of call
	☐ Urgent
	☐ Call
	☐ Returning Call
	☐ Please Call
	☐ Will Call Again
	☐ Came to see you
	☐ Wants to see you

Delivered? ☐	For	Date	Time

Caller	Position	Company

Phone	Cell	Email

Message	Type of call
	☐ Urgent
	☐ Call
	☐ Returning Call
	☐ Please Call
	☐ Will Call Again
	☐ Came to see you
	☐ Wants to see you

Delivered? ☐	For	Date	Time

Caller	Position	Company

Phone	Cell	Email

Message	Type of call
	☐ Urgent
	☐ Call
	☐ Returning Call
	☐ Please Call
	☐ Will Call Again
	☐ Came to see you
	☐ Wants to see you

Delivered? ☐	For	Date	Time

Caller	Position	Company

Phone	Cell	Email

Message	Type of call
	☐ Urgent
	☐ Call
	☐ Returning Call
	☐ Please Call
	☐ Will Call Again
	☐ Came to see you
	☐ Wants to see you

Delivered? ☐	For	Date	Time

Caller	Position	Company

Phone	Cell	Email

Message	Type of call
	☐ Urgent
	☐ Call
	☐ Returning Call
	☐ Please Call
	☐ Will Call Again
	☐ Came to see you
	☐ Wants to see you

Delivered? ☐	For	Date	Time

Caller	Position	Company

Phone	Cell	Email

Message	Type of call
	☐ Urgent
	☐ Call
	☐ Returning Call
	☐ Please Call
	☐ Will Call Again
	☐ Came to see you
	☐ Wants to see you

Delivered? ☐	For	Date	Time

Caller	Position	Company

Phone	Cell	Email

Message	Type of call
	☐ Urgent
	☐ Call
	☐ Returning Call
	☐ Please Call
	☐ Will Call Again
	☐ Came to see you
	☐ Wants to see you

Delivered?	For	Date	Time
☐			

Caller	Position	Company

Phone	Cell	Email

Message	Type of call
	☐ Urgent ☐ Call ☐ Returning Call ☐ Please Call ☐ Will Call Again ☐ Came to see you ☐ Wants to see you

Delivered?	For	Date	Time
☐			

Caller	Position	Company

Phone	Cell	Email

Message	Type of call
	☐ Urgent ☐ Call ☐ Returning Call ☐ Please Call ☐ Will Call Again ☐ Came to see you ☐ Wants to see you

Delivered?	For	Date	Time
☐			

Caller	Position	Company

Phone	Cell	Email

Message	Type of call
	☐ Urgent ☐ Call ☐ Returning Call ☐ Please Call ☐ Will Call Again ☐ Came to see you ☐ Wants to see you

Delivered? ☐	For	Date	Time

Caller	Position	Company

Phone	Cell	Email

Message	Type of call
	☐ Urgent
	☐ Call
	☐ Returning Call
	☐ Please Call
	☐ Will Call Again
	☐ Came to see you
	☐ Wants to see you

Delivered? ☐	For	Date	Time

Caller	Position	Company

Phone	Cell	Email

Message	Type of call
	☐ Urgent
	☐ Call
	☐ Returning Call
	☐ Please Call
	☐ Will Call Again
	☐ Came to see you
	☐ Wants to see you

Delivered? ☐	For	Date	Time

Caller	Position	Company

Phone	Cell	Email

Message	Type of call
	☐ Urgent
	☐ Call
	☐ Returning Call
	☐ Please Call
	☐ Will Call Again
	☐ Came to see you
	☐ Wants to see you

Delivered? ☐	For	Date	Time

Caller	Position	Company

Phone	Cell	Email

Message	Type of call
	☐ Urgent
	☐ Call
	☐ Returning Call
	☐ Please Call
	☐ Will Call Again
	☐ Came to see you
	☐ Wants to see you

Delivered? ☐	For	Date	Time

Caller	Position	Company

Phone	Cell	Email

Message	Type of call
	☐ Urgent
	☐ Call
	☐ Returning Call
	☐ Please Call
	☐ Will Call Again
	☐ Came to see you
	☐ Wants to see you

Delivered? ☐	For	Date	Time

Caller	Position	Company

Phone	Cell	Email

Message	Type of call
	☐ Urgent
	☐ Call
	☐ Returning Call
	☐ Please Call
	☐ Will Call Again
	☐ Came to see you
	☐ Wants to see you

Delivered?	For	Date	Time
☐			

Caller	Position	Company

Phone	Cell	Email

Message	Type of call
	☐ Urgent ☐ Call ☐ Returning Call ☐ Please Call ☐ Will Call Again ☐ Came to see you ☐ Wants to see you

Delivered?	For	Date	Time
☐			

Caller	Position	Company

Phone	Cell	Email

Message	Type of call
	☐ Urgent ☐ Call ☐ Returning Call ☐ Please Call ☐ Will Call Again ☐ Came to see you ☐ Wants to see you

Delivered?	For	Date	Time
☐			

Caller	Position	Company

Phone	Cell	Email

Message	Type of call
	☐ Urgent ☐ Call ☐ Returning Call ☐ Please Call ☐ Will Call Again ☐ Came to see you ☐ Wants to see you

Delivered? ☐	For	Date	Time
Caller	**Position**		**Company**
Phone	**Cell**		**Email**

Message	Type of call
	☐ Urgent ☐ Call ☐ Returning Call ☐ Please Call ☐ Will Call Again ☐ Came to see you ☐ Wants to see you

Delivered? ☐	For	Date	Time
Caller	**Position**		**Company**
Phone	**Cell**		**Email**

Message	Type of call
	☐ Urgent ☐ Call ☐ Returning Call ☐ Please Call ☐ Will Call Again ☐ Came to see you ☐ Wants to see you

Delivered? ☐	For	Date	Time
Caller	**Position**		**Company**
Phone	**Cell**		**Email**

Message	Type of call
	☐ Urgent ☐ Call ☐ Returning Call ☐ Please Call ☐ Will Call Again ☐ Came to see you ☐ Wants to see you

Delivered? ☐	For	Date	Time
Caller	Position		Company
Phone	Cell		Email

Message	Type of call
	☐ Urgent
	☐ Call
	☐ Returning Call
	☐ Please Call
	☐ Will Call Again
	☐ Came to see you
	☐ Wants to see you

Delivered? ☐	For	Date	Time
Caller	Position		Company
Phone	Cell		Email

Message	Type of call
	☐ Urgent
	☐ Call
	☐ Returning Call
	☐ Please Call
	☐ Will Call Again
	☐ Came to see you
	☐ Wants to see you

Delivered? ☐	For	Date	Time
Caller	Position		Company
Phone	Cell		Email

Message	Type of call
	☐ Urgent
	☐ Call
	☐ Returning Call
	☐ Please Call
	☐ Will Call Again
	☐ Came to see you
	☐ Wants to see you

Delivered? ☐	For	Date	Time

Caller	Position	Company

Phone	Cell	Email

Message	Type of call
	☐ Urgent
	☐ Call
	☐ Returning Call
	☐ Please Call
	☐ Will Call Again
	☐ Came to see you
	☐ Wants to see you

Delivered? ☐	For	Date	Time

Caller	Position	Company

Phone	Cell	Email

Message	Type of call
	☐ Urgent
	☐ Call
	☐ Returning Call
	☐ Please Call
	☐ Will Call Again
	☐ Came to see you
	☐ Wants to see you

Delivered? ☐	For	Date	Time

Caller	Position	Company

Phone	Cell	Email

Message	Type of call
	☐ Urgent
	☐ Call
	☐ Returning Call
	☐ Please Call
	☐ Will Call Again
	☐ Came to see you
	☐ Wants to see you

Delivered? ☐	For	Date	Time

Caller	Position	Company

Phone	Cell	Email

Message	Type of call
	☐ Urgent
	☐ Call
	☐ Returning Call
	☐ Please Call
	☐ Will Call Again
	☐ Came to see you
	☐ Wants to see you

Delivered? ☐	For	Date	Time

Caller	Position	Company

Phone	Cell	Email

Message	Type of call
	☐ Urgent
	☐ Call
	☐ Returning Call
	☐ Please Call
	☐ Will Call Again
	☐ Came to see you
	☐ Wants to see you

Delivered? ☐	For	Date	Time

Caller	Position	Company

Phone	Cell	Email

Message	Type of call
	☐ Urgent
	☐ Call
	☐ Returning Call
	☐ Please Call
	☐ Will Call Again
	☐ Came to see you
	☐ Wants to see you

Delivered? ☐	For	Date	Time

Caller	Position	Company

Phone	Cell	Email

Message	Type of call
	☐ Urgent ☐ Call ☐ Returning Call ☐ Please Call ☐ Will Call Again ☐ Came to see you ☐ Wants to see you

Delivered? ☐	For	Date	Time

Caller	Position	Company

Phone	Cell	Email

Message	Type of call
	☐ Urgent ☐ Call ☐ Returning Call ☐ Please Call ☐ Will Call Again ☐ Came to see you ☐ Wants to see you

Delivered? ☐	For	Date	Time

Caller	Position	Company

Phone	Cell	Email

Message	Type of call
	☐ Urgent ☐ Call ☐ Returning Call ☐ Please Call ☐ Will Call Again ☐ Came to see you ☐ Wants to see you

Delivered? ☐	For	Date	Time

Caller	Position	Company

Phone	Cell	Email

Message	Type of call
	☐ Urgent
	☐ Call
	☐ Returning Call
	☐ Please Call
	☐ Will Call Again
	☐ Came to see you
	☐ Wants to see you

Delivered? ☐	For	Date	Time

Caller	Position	Company

Phone	Cell	Email

Message	Type of call
	☐ Urgent
	☐ Call
	☐ Returning Call
	☐ Please Call
	☐ Will Call Again
	☐ Came to see you
	☐ Wants to see you

Delivered? ☐	For	Date	Time

Caller	Position	Company

Phone	Cell	Email

Message	Type of call
	☐ Urgent
	☐ Call
	☐ Returning Call
	☐ Please Call
	☐ Will Call Again
	☐ Came to see you
	☐ Wants to see you

Delivered? ☐	For	Date	Time

Caller	Position	Company

Phone	Cell	Email

Message	Type of call
	☐ Urgent
	☐ Call
	☐ Returning Call
	☐ Please Call
	☐ Will Call Again
	☐ Came to see you
	☐ Wants to see you

Delivered? ☐	For	Date	Time

Caller	Position	Company

Phone	Cell	Email

Message	Type of call
	☐ Urgent
	☐ Call
	☐ Returning Call
	☐ Please Call
	☐ Will Call Again
	☐ Came to see you
	☐ Wants to see you

Delivered? ☐	For	Date	Time

Caller	Position	Company

Phone	Cell	Email

Message	Type of call
	☐ Urgent
	☐ Call
	☐ Returning Call
	☐ Please Call
	☐ Will Call Again
	☐ Came to see you
	☐ Wants to see you

Delivered? ☐	For	Date	Time

Caller	Position	Company

Phone	Cell	Email

Message	Type of call
	☐ Urgent
	☐ Call
	☐ Returning Call
	☐ Please Call
	☐ Will Call Again
	☐ Came to see you
	☐ Wants to see you

Delivered? ☐	For	Date	Time

Caller	Position	Company

Phone	Cell	Email

Message	Type of call
	☐ Urgent
	☐ Call
	☐ Returning Call
	☐ Please Call
	☐ Will Call Again
	☐ Came to see you
	☐ Wants to see you

Delivered? ☐	For	Date	Time

Caller	Position	Company

Phone	Cell	Email

Message	Type of call
	☐ Urgent
	☐ Call
	☐ Returning Call
	☐ Please Call
	☐ Will Call Again
	☐ Came to see you
	☐ Wants to see you

Delivered? ☐	For	Date	Time

Caller	Position	Company

Phone	Cell	Email

Message	Type of call
	☐ Urgent ☐ Call ☐ Returning Call ☐ Please Call ☐ Will Call Again ☐ Came to see you ☐ Wants to see you

Delivered? ☐	For	Date	Time

Caller	Position	Company

Phone	Cell	Email

Message	Type of call
	☐ Urgent ☐ Call ☐ Returning Call ☐ Please Call ☐ Will Call Again ☐ Came to see you ☐ Wants to see you

Delivered? ☐	For	Date	Time

Caller	Position	Company

Phone	Cell	Email

Message	Type of call
	☐ Urgent ☐ Call ☐ Returning Call ☐ Please Call ☐ Will Call Again ☐ Came to see you ☐ Wants to see you

Delivered? ☐	For	Date	Time

Caller	Position	Company

Phone	Cell	Email

Message	Type of call
	☐ Urgent
	☐ Call
	☐ Returning Call
	☐ Please Call
	☐ Will Call Again
	☐ Came to see you
	☐ Wants to see you

Delivered? ☐	For	Date	Time

Caller	Position	Company

Phone	Cell	Email

Message	Type of call
	☐ Urgent
	☐ Call
	☐ Returning Call
	☐ Please Call
	☐ Will Call Again
	☐ Came to see you
	☐ Wants to see you

Delivered? ☐	For	Date	Time

Caller	Position	Company

Phone	Cell	Email

Message	Type of call
	☐ Urgent
	☐ Call
	☐ Returning Call
	☐ Please Call
	☐ Will Call Again
	☐ Came to see you
	☐ Wants to see you

Delivered? ☐	For	Date	Time

Caller	Position	Company

Phone	Cell	Email

Message	Type of call
	☐ Urgent
	☐ Call
	☐ Returning Call
	☐ Please Call
	☐ Will Call Again
	☐ Came to see you
	☐ Wants to see you

Delivered? ☐	For	Date	Time

Caller	Position	Company

Phone	Cell	Email

Message	Type of call
	☐ Urgent
	☐ Call
	☐ Returning Call
	☐ Please Call
	☐ Will Call Again
	☐ Came to see you
	☐ Wants to see you

Delivered? ☐	For	Date	Time

Caller	Position	Company

Phone	Cell	Email

Message	Type of call
	☐ Urgent
	☐ Call
	☐ Returning Call
	☐ Please Call
	☐ Will Call Again
	☐ Came to see you
	☐ Wants to see you

Delivered?	For	Date	Time
☐			

Caller	Position	Company

Phone	Cell	Email

Message	Type of call
	☐ Urgent ☐ Call ☐ Returning Call ☐ Please Call ☐ Will Call Again ☐ Came to see you ☐ Wants to see you

Delivered?	For	Date	Time
☐			

Caller	Position	Company

Phone	Cell	Email

Message	Type of call
	☐ Urgent ☐ Call ☐ Returning Call ☐ Please Call ☐ Will Call Again ☐ Came to see you ☐ Wants to see you

Delivered?	For	Date	Time
☐			

Caller	Position	Company

Phone	Cell	Email

Message	Type of call
	☐ Urgent ☐ Call ☐ Returning Call ☐ Please Call ☐ Will Call Again ☐ Came to see you ☐ Wants to see you

Delivered? ☐	For	Date	Time

Caller	Position	Company

Phone	Cell	Email

Message	Type of call
	☐ Urgent
	☐ Call
	☐ Returning Call
	☐ Please Call
	☐ Will Call Again
	☐ Came to see you
	☐ Wants to see you

Delivered? ☐	For	Date	Time

Caller	Position	Company

Phone	Cell	Email

Message	Type of call
	☐ Urgent
	☐ Call
	☐ Returning Call
	☐ Please Call
	☐ Will Call Again
	☐ Came to see you
	☐ Wants to see you

Delivered? ☐	For	Date	Time

Caller	Position	Company

Phone	Cell	Email

Message	Type of call
	☐ Urgent
	☐ Call
	☐ Returning Call
	☐ Please Call
	☐ Will Call Again
	☐ Came to see you
	☐ Wants to see you

Delivered? ☐	For	Date	Time

Caller	Position	Company

Phone	Cell	Email

Message	Type of call
	☐ Urgent
	☐ Call
	☐ Returning Call
	☐ Please Call
	☐ Will Call Again
	☐ Came to see you
	☐ Wants to see you

Delivered? ☐	For	Date	Time

Caller	Position	Company

Phone	Cell	Email

Message	Type of call
	☐ Urgent
	☐ Call
	☐ Returning Call
	☐ Please Call
	☐ Will Call Again
	☐ Came to see you
	☐ Wants to see you

Delivered? ☐	For	Date	Time

Caller	Position	Company

Phone	Cell	Email

Message	Type of call
	☐ Urgent
	☐ Call
	☐ Returning Call
	☐ Please Call
	☐ Will Call Again
	☐ Came to see you
	☐ Wants to see you

Delivered? ☐	For	Date	Time

Caller	Position	Company

Phone	Cell	Email

Message	Type of call
	☐ Urgent ☐ Call ☐ Returning Call ☐ Please Call ☐ Will Call Again ☐ Came to see you ☐ Wants to see you

Delivered? ☐	For	Date	Time

Caller	Position	Company

Phone	Cell	Email

Message	Type of call
	☐ Urgent ☐ Call ☐ Returning Call ☐ Please Call ☐ Will Call Again ☐ Came to see you ☐ Wants to see you

Delivered? ☐	For	Date	Time

Caller	Position	Company

Phone	Cell	Email

Message	Type of call
	☐ Urgent ☐ Call ☐ Returning Call ☐ Please Call ☐ Will Call Again ☐ Came to see you ☐ Wants to see you

Delivered? ☐	For	Date	Time

Caller	Position	Company

Phone	Cell	Email

Message	Type of call
	☐ Urgent
	☐ Call
	☐ Returning Call
	☐ Please Call
	☐ Will Call Again
	☐ Came to see you
	☐ Wants to see you

Delivered? ☐	For	Date	Time

Caller	Position	Company

Phone	Cell	Email

Message	Type of call
	☐ Urgent
	☐ Call
	☐ Returning Call
	☐ Please Call
	☐ Will Call Again
	☐ Came to see you
	☐ Wants to see you

Delivered? ☐	For	Date	Time

Caller	Position	Company

Phone	Cell	Email

Message	Type of call
	☐ Urgent
	☐ Call
	☐ Returning Call
	☐ Please Call
	☐ Will Call Again
	☐ Came to see you
	☐ Wants to see you

Delivered? ☐	For	Date	Time

Caller	Position	Company

Phone	Cell	Email

Message	Type of call
	☐ Urgent ☐ Call ☐ Returning Call ☐ Please Call ☐ Will Call Again ☐ Came to see you ☐ Wants to see you

Delivered? ☐	For	Date	Time

Caller	Position	Company

Phone	Cell	Email

Message	Type of call
	☐ Urgent ☐ Call ☐ Returning Call ☐ Please Call ☐ Will Call Again ☐ Came to see you ☐ Wants to see you

Delivered? ☐	For	Date	Time

Caller	Position	Company

Phone	Cell	Email

Message	Type of call
	☐ Urgent ☐ Call ☐ Returning Call ☐ Please Call ☐ Will Call Again ☐ Came to see you ☐ Wants to see you

Delivered? ☐	For	Date	Time

Caller	Position	Company

Phone	Cell	Email

Message	Type of call
	☐ Urgent
	☐ Call
	☐ Returning Call
	☐ Please Call
	☐ Will Call Again
	☐ Came to see you
	☐ Wants to see you

Delivered? ☐	For	Date	Time

Caller	Position	Company

Phone	Cell	Email

Message	Type of call
	☐ Urgent
	☐ Call
	☐ Returning Call
	☐ Please Call
	☐ Will Call Again
	☐ Came to see you
	☐ Wants to see you

Delivered? ☐	For	Date	Time

Caller	Position	Company

Phone	Cell	Email

Message	Type of call
	☐ Urgent
	☐ Call
	☐ Returning Call
	☐ Please Call
	☐ Will Call Again
	☐ Came to see you
	☐ Wants to see you

Delivered? ☐	For	Date	Time

Caller	Position	Company

Phone	Cell	Email

Message	Type of call
	☐ Urgent
	☐ Call
	☐ Returning Call
	☐ Please Call
	☐ Will Call Again
	☐ Came to see you
	☐ Wants to see you

Delivered? ☐	For	Date	Time

Caller	Position	Company

Phone	Cell	Email

Message	Type of call
	☐ Urgent
	☐ Call
	☐ Returning Call
	☐ Please Call
	☐ Will Call Again
	☐ Came to see you
	☐ Wants to see you

Delivered? ☐	For	Date	Time

Caller	Position	Company

Phone	Cell	Email

Message	Type of call
	☐ Urgent
	☐ Call
	☐ Returning Call
	☐ Please Call
	☐ Will Call Again
	☐ Came to see you
	☐ Wants to see you

Delivered? ☐	For	Date	Time

Caller	Position	Company

Phone	Cell	Email

Message	Type of call
	☐ Urgent
	☐ Call
	☐ Returning Call
	☐ Please Call
	☐ Will Call Again
	☐ Came to see you
	☐ Wants to see you

Delivered? ☐	For	Date	Time

Caller	Position	Company

Phone	Cell	Email

Message	Type of call
	☐ Urgent
	☐ Call
	☐ Returning Call
	☐ Please Call
	☐ Will Call Again
	☐ Came to see you
	☐ Wants to see you

Delivered? ☐	For	Date	Time

Caller	Position	Company

Phone	Cell	Email

Message	Type of call
	☐ Urgent
	☐ Call
	☐ Returning Call
	☐ Please Call
	☐ Will Call Again
	☐ Came to see you
	☐ Wants to see you

Delivered? ☐	For	Date	Time

Caller	Position	Company

Phone	Cell	Email

Message	Type of call
	☐ Urgent
	☐ Call
	☐ Returning Call
	☐ Please Call
	☐ Will Call Again
	☐ Came to see you
	☐ Wants to see you

Delivered? ☐	For	Date	Time

Caller	Position	Company

Phone	Cell	Email

Message	Type of call
	☐ Urgent
	☐ Call
	☐ Returning Call
	☐ Please Call
	☐ Will Call Again
	☐ Came to see you
	☐ Wants to see you

Delivered? ☐	For	Date	Time

Caller	Position	Company

Phone	Cell	Email

Message	Type of call
	☐ Urgent
	☐ Call
	☐ Returning Call
	☐ Please Call
	☐ Will Call Again
	☐ Came to see you
	☐ Wants to see you

Delivered? ☐	For	Date	Time
Caller	Position		Company
Phone	Cell		Email

Message	Type of call
	☐ Urgent ☐ Call ☐ Returning Call ☐ Please Call ☐ Will Call Again ☐ Came to see you ☐ Wants to see you

Delivered? ☐	For	Date	Time
Caller	Position		Company
Phone	Cell		Email

Message	Type of call
	☐ Urgent ☐ Call ☐ Returning Call ☐ Please Call ☐ Will Call Again ☐ Came to see you ☐ Wants to see you

Delivered? ☐	For	Date	Time
Caller	Position		Company
Phone	Cell		Email

Message	Type of call
	☐ Urgent ☐ Call ☐ Returning Call ☐ Please Call ☐ Will Call Again ☐ Came to see you ☐ Wants to see you

Delivered?	For	Date	Time
☐			

Caller	Position	Company

Phone	Cell	Email

Message	Type of call
	☐ Urgent ☐ Call ☐ Returning Call ☐ Please Call ☐ Will Call Again ☐ Came to see you ☐ Wants to see you

Delivered?	For	Date	Time
☐			

Caller	Position	Company

Phone	Cell	Email

Message	Type of call
	☐ Urgent ☐ Call ☐ Returning Call ☐ Please Call ☐ Will Call Again ☐ Came to see you ☐ Wants to see you

Delivered?	For	Date	Time
☐			

Caller	Position	Company

Phone	Cell	Email

Message	Type of call
	☐ Urgent ☐ Call ☐ Returning Call ☐ Please Call ☐ Will Call Again ☐ Came to see you ☐ Wants to see you

Delivered? ☐	For	Date	Time

Caller	Position	Company

Phone	Cell	Email

Message	Type of call
	☐ Urgent
	☐ Call
	☐ Returning Call
	☐ Please Call
	☐ Will Call Again
	☐ Came to see you
	☐ Wants to see you

Delivered? ☐	For	Date	Time

Caller	Position	Company

Phone	Cell	Email

Message	Type of call
	☐ Urgent
	☐ Call
	☐ Returning Call
	☐ Please Call
	☐ Will Call Again
	☐ Came to see you
	☐ Wants to see you

Delivered? ☐	For	Date	Time

Caller	Position	Company

Phone	Cell	Email

Message	Type of call
	☐ Urgent
	☐ Call
	☐ Returning Call
	☐ Please Call
	☐ Will Call Again
	☐ Came to see you
	☐ Wants to see you

Delivered? ☐	For	Date	Time

Caller	Position	Company

Phone	Cell	Email

Message	Type of call
	☐ Urgent ☐ Call ☐ Returning Call ☐ Please Call ☐ Will Call Again ☐ Came to see you ☐ Wants to see you

Delivered? ☐	For	Date	Time

Caller	Position	Company

Phone	Cell	Email

Message	Type of call
	☐ Urgent ☐ Call ☐ Returning Call ☐ Please Call ☐ Will Call Again ☐ Came to see you ☐ Wants to see you

Delivered? ☐	For	Date	Time

Caller	Position	Company

Phone	Cell	Email

Message	Type of call
	☐ Urgent ☐ Call ☐ Returning Call ☐ Please Call ☐ Will Call Again ☐ Came to see you ☐ Wants to see you

Delivered? ☐	For	Date	Time

Caller	Position	Company

Phone	Cell	Email

Message	Type of call
	☐ Urgent
	☐ Call
	☐ Returning Call
	☐ Please Call
	☐ Will Call Again
	☐ Came to see you
	☐ Wants to see you

Delivered? ☐	For	Date	Time

Caller	Position	Company

Phone	Cell	Email

Message	Type of call
	☐ Urgent
	☐ Call
	☐ Returning Call
	☐ Please Call
	☐ Will Call Again
	☐ Came to see you
	☐ Wants to see you

Delivered? ☐	For	Date	Time

Caller	Position	Company

Phone	Cell	Email

Message	Type of call
	☐ Urgent
	☐ Call
	☐ Returning Call
	☐ Please Call
	☐ Will Call Again
	☐ Came to see you
	☐ Wants to see you

Delivered? ☐	For	Date	Time

Caller	Position	Company

Phone	Cell	Email

Message	Type of call
	☐ Urgent ☐ Call ☐ Returning Call ☐ Please Call ☐ Will Call Again ☐ Came to see you ☐ Wants to see you

Delivered? ☐	For	Date	Time

Caller	Position	Company

Phone	Cell	Email

Message	Type of call
	☐ Urgent ☐ Call ☐ Returning Call ☐ Please Call ☐ Will Call Again ☐ Came to see you ☐ Wants to see you

Delivered? ☐	For	Date	Time

Caller	Position	Company

Phone	Cell	Email

Message	Type of call
	☐ Urgent ☐ Call ☐ Returning Call ☐ Please Call ☐ Will Call Again ☐ Came to see you ☐ Wants to see you

Delivered? ☐	For	Date	Time

Caller	Position	Company

Phone	Cell	Email

Message	Type of call
	☐ Urgent
	☐ Call
	☐ Returning Call
	☐ Please Call
	☐ Will Call Again
	☐ Came to see you
	☐ Wants to see you

Delivered? ☐	For	Date	Time

Caller	Position	Company

Phone	Cell	Email

Message	Type of call
	☐ Urgent
	☐ Call
	☐ Returning Call
	☐ Please Call
	☐ Will Call Again
	☐ Came to see you
	☐ Wants to see you

Delivered? ☐	For	Date	Time

Caller	Position	Company

Phone	Cell	Email

Message	Type of call
	☐ Urgent
	☐ Call
	☐ Returning Call
	☐ Please Call
	☐ Will Call Again
	☐ Came to see you
	☐ Wants to see you

Delivered? ☐	For	Date	Time

Caller	Position	Company

Phone	Cell	Email

Message	Type of call
	☐ Urgent
	☐ Call
	☐ Returning Call
	☐ Please Call
	☐ Will Call Again
	☐ Came to see you
	☐ Wants to see you

Delivered? ☐	For	Date	Time

Caller	Position	Company

Phone	Cell	Email

Message	Type of call
	☐ Urgent
	☐ Call
	☐ Returning Call
	☐ Please Call
	☐ Will Call Again
	☐ Came to see you
	☐ Wants to see you

Delivered? ☐	For	Date	Time

Caller	Position	Company

Phone	Cell	Email

Message	Type of call
	☐ Urgent
	☐ Call
	☐ Returning Call
	☐ Please Call
	☐ Will Call Again
	☐ Came to see you
	☐ Wants to see you

Delivered? ☐	For	Date	Time

Caller	Position	Company

Phone	Cell	Email

Message	Type of call
	☐ Urgent
	☐ Call
	☐ Returning Call
	☐ Please Call
	☐ Will Call Again
	☐ Came to see you
	☐ Wants to see you

Delivered? ☐	For	Date	Time

Caller	Position	Company

Phone	Cell	Email

Message	Type of call
	☐ Urgent
	☐ Call
	☐ Returning Call
	☐ Please Call
	☐ Will Call Again
	☐ Came to see you
	☐ Wants to see you

Delivered? ☐	For	Date	Time

Caller	Position	Company

Phone	Cell	Email

Message	Type of call
	☐ Urgent
	☐ Call
	☐ Returning Call
	☐ Please Call
	☐ Will Call Again
	☐ Came to see you
	☐ Wants to see you

Delivered? ☐	For	Date	Time

Caller	Position	Company

Phone	Cell	Email

Message	Type of call
	☐ Urgent ☐ Call ☐ Returning Call ☐ Please Call ☐ Will Call Again ☐ Came to see you ☐ Wants to see you

Delivered? ☐	For	Date	Time

Caller	Position	Company

Phone	Cell	Email

Message	Type of call
	☐ Urgent ☐ Call ☐ Returning Call ☐ Please Call ☐ Will Call Again ☐ Came to see you ☐ Wants to see you

Delivered? ☐	For	Date	Time

Caller	Position	Company

Phone	Cell	Email

Message	Type of call
	☐ Urgent ☐ Call ☐ Returning Call ☐ Please Call ☐ Will Call Again ☐ Came to see you ☐ Wants to see you

Delivered? ☐	For	Date	Time

Caller	Position	Company

Phone	Cell	Email

Message	Type of call
	☐ Urgent
	☐ Call
	☐ Returning Call
	☐ Please Call
	☐ Will Call Again
	☐ Came to see you
	☐ Wants to see you

Delivered? ☐	For	Date	Time

Caller	Position	Company

Phone	Cell	Email

Message	Type of call
	☐ Urgent
	☐ Call
	☐ Returning Call
	☐ Please Call
	☐ Will Call Again
	☐ Came to see you
	☐ Wants to see you

Delivered? ☐	For	Date	Time

Caller	Position	Company

Phone	Cell	Email

Message	Type of call
	☐ Urgent
	☐ Call
	☐ Returning Call
	☐ Please Call
	☐ Will Call Again
	☐ Came to see you
	☐ Wants to see you

Delivered? ☐	For	Date	Time

Caller	Position	Company

Phone	Cell	Email

Message	Type of call
	☐ Urgent ☐ Call ☐ Returning Call ☐ Please Call ☐ Will Call Again ☐ Came to see you ☐ Wants to see you

Delivered? ☐	For	Date	Time

Caller	Position	Company

Phone	Cell	Email

Message	Type of call
	☐ Urgent ☐ Call ☐ Returning Call ☐ Please Call ☐ Will Call Again ☐ Came to see you ☐ Wants to see you

Delivered? ☐	For	Date	Time

Caller	Position	Company

Phone	Cell	Email

Message	Type of call
	☐ Urgent ☐ Call ☐ Returning Call ☐ Please Call ☐ Will Call Again ☐ Came to see you ☐ Wants to see you

Delivered? ☐	For	Date	Time

Caller	Position	Company

Phone	Cell	Email

Message	Type of call
	☐ Urgent
	☐ Call
	☐ Returning Call
	☐ Please Call
	☐ Will Call Again
	☐ Came to see you
	☐ Wants to see you

Delivered? ☐	For	Date	Time

Caller	Position	Company

Phone	Cell	Email

Message	Type of call
	☐ Urgent
	☐ Call
	☐ Returning Call
	☐ Please Call
	☐ Will Call Again
	☐ Came to see you
	☐ Wants to see you

Delivered? ☐	For	Date	Time

Caller	Position	Company

Phone	Cell	Email

Message	Type of call
	☐ Urgent
	☐ Call
	☐ Returning Call
	☐ Please Call
	☐ Will Call Again
	☐ Came to see you
	☐ Wants to see you

Delivered? ☐	For	Date	Time

Caller	Position	Company

Phone	Cell	Email

Message	Type of call
	☐ Urgent
	☐ Call
	☐ Returning Call
	☐ Please Call
	☐ Will Call Again
	☐ Came to see you
	☐ Wants to see you

Delivered? ☐	For	Date	Time

Caller	Position	Company

Phone	Cell	Email

Message	Type of call
	☐ Urgent
	☐ Call
	☐ Returning Call
	☐ Please Call
	☐ Will Call Again
	☐ Came to see you
	☐ Wants to see you

Delivered? ☐	For	Date	Time

Caller	Position	Company

Phone	Cell	Email

Message	Type of call
	☐ Urgent
	☐ Call
	☐ Returning Call
	☐ Please Call
	☐ Will Call Again
	☐ Came to see you
	☐ Wants to see you

Delivered? ☐	For	Date	Time

Caller	Position	Company

Phone	Cell	Email

Message	Type of call
	☐ Urgent
	☐ Call
	☐ Returning Call
	☐ Please Call
	☐ Will Call Again
	☐ Came to see you
	☐ Wants to see you

Delivered? ☐	For	Date	Time

Caller	Position	Company

Phone	Cell	Email

Message	Type of call
	☐ Urgent
	☐ Call
	☐ Returning Call
	☐ Please Call
	☐ Will Call Again
	☐ Came to see you
	☐ Wants to see you

Delivered? ☐	For	Date	Time

Caller	Position	Company

Phone	Cell	Email

Message	Type of call
	☐ Urgent
	☐ Call
	☐ Returning Call
	☐ Please Call
	☐ Will Call Again
	☐ Came to see you
	☐ Wants to see you

Delivered? ☐	For	Date	Time

Caller	Position	Company

Phone	Cell	Email

Message	Type of call
	☐ Urgent
	☐ Call
	☐ Returning Call
	☐ Please Call
	☐ Will Call Again
	☐ Came to see you
	☐ Wants to see you

Delivered? ☐	For	Date	Time

Caller	Position	Company

Phone	Cell	Email

Message	Type of call
	☐ Urgent
	☐ Call
	☐ Returning Call
	☐ Please Call
	☐ Will Call Again
	☐ Came to see you
	☐ Wants to see you

Delivered? ☐	For	Date	Time

Caller	Position	Company

Phone	Cell	Email

Message	Type of call
	☐ Urgent
	☐ Call
	☐ Returning Call
	☐ Please Call
	☐ Will Call Again
	☐ Came to see you
	☐ Wants to see you

Delivered? ☐	For	Date	Time

Caller	Position	Company

Phone	Cell	Email

Message	Type of call
	☐ Urgent
	☐ Call
	☐ Returning Call
	☐ Please Call
	☐ Will Call Again
	☐ Came to see you
	☐ Wants to see you

Delivered? ☐	For	Date	Time

Caller	Position	Company

Phone	Cell	Email

Message	Type of call
	☐ Urgent
	☐ Call
	☐ Returning Call
	☐ Please Call
	☐ Will Call Again
	☐ Came to see you
	☐ Wants to see you

Delivered? ☐	For	Date	Time

Caller	Position	Company

Phone	Cell	Email

Message	Type of call
	☐ Urgent
	☐ Call
	☐ Returning Call
	☐ Please Call
	☐ Will Call Again
	☐ Came to see you
	☐ Wants to see you

Delivered? ☐	For	Date	Time

Caller	Position	Company

Phone	Cell	Email

Message	Type of call
	☐ Urgent
	☐ Call
	☐ Returning Call
	☐ Please Call
	☐ Will Call Again
	☐ Came to see you
	☐ Wants to see you

Delivered? ☐	For	Date	Time

Caller	Position	Company

Phone	Cell	Email

Message	Type of call
	☐ Urgent
	☐ Call
	☐ Returning Call
	☐ Please Call
	☐ Will Call Again
	☐ Came to see you
	☐ Wants to see you

Delivered? ☐	For	Date	Time

Caller	Position	Company

Phone	Cell	Email

Message	Type of call
	☐ Urgent
	☐ Call
	☐ Returning Call
	☐ Please Call
	☐ Will Call Again
	☐ Came to see you
	☐ Wants to see you

Delivered? ☐	For	Date	Time

Caller	Position	Company

Phone	Cell	Email

Message	Type of call
	☐ Urgent
	☐ Call
	☐ Returning Call
	☐ Please Call
	☐ Will Call Again
	☐ Came to see you
	☐ Wants to see you

Delivered? ☐	For	Date	Time

Caller	Position	Company

Phone	Cell	Email

Message	Type of call
	☐ Urgent
	☐ Call
	☐ Returning Call
	☐ Please Call
	☐ Will Call Again
	☐ Came to see you
	☐ Wants to see you

Delivered? ☐	For	Date	Time

Caller	Position	Company

Phone	Cell	Email

Message	Type of call
	☐ Urgent
	☐ Call
	☐ Returning Call
	☐ Please Call
	☐ Will Call Again
	☐ Came to see you
	☐ Wants to see you

Delivered? ☐	For	Date	Time

Caller	Position	Company

Phone	Cell	Email

Message	Type of call
	☐ Urgent
	☐ Call
	☐ Returning Call
	☐ Please Call
	☐ Will Call Again
	☐ Came to see you
	☐ Wants to see you

Delivered? ☐	For	Date	Time

Caller	Position	Company

Phone	Cell	Email

Message	Type of call
	☐ Urgent
	☐ Call
	☐ Returning Call
	☐ Please Call
	☐ Will Call Again
	☐ Came to see you
	☐ Wants to see you

Delivered? ☐	For	Date	Time

Caller	Position	Company

Phone	Cell	Email

Message	Type of call
	☐ Urgent
	☐ Call
	☐ Returning Call
	☐ Please Call
	☐ Will Call Again
	☐ Came to see you
	☐ Wants to see you

Delivered? ☐	For	Date	Time

Caller	Position	Company

Phone	Cell	Email

Message	Type of call
	☐ Urgent
	☐ Call
	☐ Returning Call
	☐ Please Call
	☐ Will Call Again
	☐ Came to see you
	☐ Wants to see you

Delivered? ☐	For	Date	Time

Caller	Position	Company

Phone	Cell	Email

Message	Type of call
	☐ Urgent
	☐ Call
	☐ Returning Call
	☐ Please Call
	☐ Will Call Again
	☐ Came to see you
	☐ Wants to see you

Delivered? ☐	For	Date	Time

Caller	Position	Company

Phone	Cell	Email

Message	Type of call
	☐ Urgent
	☐ Call
	☐ Returning Call
	☐ Please Call
	☐ Will Call Again
	☐ Came to see you
	☐ Wants to see you

Delivered? ☐	For	Date	Time

Caller	Position	Company

Phone	Cell	Email

Message	Type of call
	☐ Urgent
	☐ Call
	☐ Returning Call
	☐ Please Call
	☐ Will Call Again
	☐ Came to see you
	☐ Wants to see you

Delivered? ☐	For	Date	Time

Caller	Position	Company

Phone	Cell	Email

Message	Type of call
	☐ Urgent
	☐ Call
	☐ Returning Call
	☐ Please Call
	☐ Will Call Again
	☐ Came to see you
	☐ Wants to see you

Delivered? ☐	For	Date	Time

Caller	Position	Company

Phone	Cell	Email

Message	Type of call
	☐ Urgent
	☐ Call
	☐ Returning Call
	☐ Please Call
	☐ Will Call Again
	☐ Came to see you
	☐ Wants to see you

Delivered? ☐	For	Date	Time

Caller	Position	Company

Phone	Cell	Email

Message	Type of call
	☐ Urgent
	☐ Call
	☐ Returning Call
	☐ Please Call
	☐ Will Call Again
	☐ Came to see you
	☐ Wants to see you

Delivered? ☐	For	Date	Time

Caller	Position	Company

Phone	Cell	Email

Message	Type of call
	☐ Urgent
	☐ Call
	☐ Returning Call
	☐ Please Call
	☐ Will Call Again
	☐ Came to see you
	☐ Wants to see you

Delivered? ☐	For	Date	Time

Caller	Position	Company

Phone	Cell	Email

Message	Type of call
	☐ Urgent
	☐ Call
	☐ Returning Call
	☐ Please Call
	☐ Will Call Again
	☐ Came to see you
	☐ Wants to see you

Delivered? ☐	For	Date	Time

Caller	Position	Company

Phone	Cell	Email

Message	Type of call
	☐ Urgent
	☐ Call
	☐ Returning Call
	☐ Please Call
	☐ Will Call Again
	☐ Came to see you
	☐ Wants to see you

Delivered? ☐	For	Date	Time

Caller	Position	Company

Phone	Cell	Email

Message	Type of call
	☐ Urgent
	☐ Call
	☐ Returning Call
	☐ Please Call
	☐ Will Call Again
	☐ Came to see you
	☐ Wants to see you

Delivered? ☐	For	Date	Time

Caller	Position	Company

Phone	Cell	Email

Message	Type of call
	☐ Urgent
	☐ Call
	☐ Returning Call
	☐ Please Call
	☐ Will Call Again
	☐ Came to see you
	☐ Wants to see you

Delivered? ☐	For	Date	Time

Caller	Position	Company

Phone	Cell	Email

Message	Type of call
	☐ Urgent
	☐ Call
	☐ Returning Call
	☐ Please Call
	☐ Will Call Again
	☐ Came to see you
	☐ Wants to see you

Delivered? ☐	For	Date	Time

Caller	Position	Company

Phone	Cell	Email

Message	Type of call
	☐ Urgent
	☐ Call
	☐ Returning Call
	☐ Please Call
	☐ Will Call Again
	☐ Came to see you
	☐ Wants to see you

Delivered? ☐	For	Date	Time

Caller	Position	Company

Phone	Cell	Email

Message	Type of call
	☐ Urgent
	☐ Call
	☐ Returning Call
	☐ Please Call
	☐ Will Call Again
	☐ Came to see you
	☐ Wants to see you

Delivered? ☐	For	Date	Time

Caller	Position	Company

Phone	Cell	Email

Message	Type of call
	☐ Urgent
	☐ Call
	☐ Returning Call
	☐ Please Call
	☐ Will Call Again
	☐ Came to see you
	☐ Wants to see you

Delivered? ☐	For	Date	Time

Caller	Position	Company

Phone	Cell	Email

Message	Type of call
	☐ Urgent
	☐ Call
	☐ Returning Call
	☐ Please Call
	☐ Will Call Again
	☐ Came to see you
	☐ Wants to see you

Delivered? ☐	For	Date	Time

Caller	Position	Company

Phone	Cell	Email

Message	Type of call
	☐ Urgent
	☐ Call
	☐ Returning Call
	☐ Please Call
	☐ Will Call Again
	☐ Came to see you
	☐ Wants to see you

Delivered? ☐	For	Date	Time
Caller	Position		Company
Phone	Cell		Email

Message	Type of call
	☐ Urgent
	☐ Call
	☐ Returning Call
	☐ Please Call
	☐ Will Call Again
	☐ Came to see you
	☐ Wants to see you

Delivered? ☐	For	Date	Time
Caller	Position		Company
Phone	Cell		Email

Message	Type of call
	☐ Urgent
	☐ Call
	☐ Returning Call
	☐ Please Call
	☐ Will Call Again
	☐ Came to see you
	☐ Wants to see you

Delivered? ☐	For	Date	Time
Caller	Position		Company
Phone	Cell		Email

Message	Type of call
	☐ Urgent
	☐ Call
	☐ Returning Call
	☐ Please Call
	☐ Will Call Again
	☐ Came to see you
	☐ Wants to see you

Delivered? ☐	For	Date	Time

Caller	Position	Company

Phone	Cell	Email

Message	Type of call
	☐ Urgent
	☐ Call
	☐ Returning Call
	☐ Please Call
	☐ Will Call Again
	☐ Came to see you
	☐ Wants to see you

Delivered? ☐	For	Date	Time

Caller	Position	Company

Phone	Cell	Email

Message	Type of call
	☐ Urgent
	☐ Call
	☐ Returning Call
	☐ Please Call
	☐ Will Call Again
	☐ Came to see you
	☐ Wants to see you

Delivered? ☐	For	Date	Time

Caller	Position	Company

Phone	Cell	Email

Message	Type of call
	☐ Urgent
	☐ Call
	☐ Returning Call
	☐ Please Call
	☐ Will Call Again
	☐ Came to see you
	☐ Wants to see you

Delivered? ☐	For	Date	Time

Caller	Position	Company

Phone	Cell	Email

Message	Type of call
	☐ Urgent
	☐ Call
	☐ Returning Call
	☐ Please Call
	☐ Will Call Again
	☐ Came to see you
	☐ Wants to see you

Delivered? ☐	For	Date	Time

Caller	Position	Company

Phone	Cell	Email

Message	Type of call
	☐ Urgent
	☐ Call
	☐ Returning Call
	☐ Please Call
	☐ Will Call Again
	☐ Came to see you
	☐ Wants to see you

Delivered? ☐	For	Date	Time

Caller	Position	Company

Phone	Cell	Email

Message	Type of call
	☐ Urgent
	☐ Call
	☐ Returning Call
	☐ Please Call
	☐ Will Call Again
	☐ Came to see you
	☐ Wants to see you

Delivered? ☐	For	Date	Time

Caller	Position	Company

Phone	Cell	Email

Message	Type of call
	☐ Urgent
	☐ Call
	☐ Returning Call
	☐ Please Call
	☐ Will Call Again
	☐ Came to see you
	☐ Wants to see you

Delivered? ☐	For	Date	Time

Caller	Position	Company

Phone	Cell	Email

Message	Type of call
	☐ Urgent
	☐ Call
	☐ Returning Call
	☐ Please Call
	☐ Will Call Again
	☐ Came to see you
	☐ Wants to see you

Delivered? ☐	For	Date	Time

Caller	Position	Company

Phone	Cell	Email

Message	Type of call
	☐ Urgent
	☐ Call
	☐ Returning Call
	☐ Please Call
	☐ Will Call Again
	☐ Came to see you
	☐ Wants to see you

Delivered? ☐	For	Date	Time

Caller	Position	Company

Phone	Cell	Email

Message	Type of call
	☐ Urgent
	☐ Call
	☐ Returning Call
	☐ Please Call
	☐ Will Call Again
	☐ Came to see you
	☐ Wants to see you

Delivered? ☐	For	Date	Time

Caller	Position	Company

Phone	Cell	Email

Message	Type of call
	☐ Urgent
	☐ Call
	☐ Returning Call
	☐ Please Call
	☐ Will Call Again
	☐ Came to see you
	☐ Wants to see you

Delivered? ☐	For	Date	Time

Caller	Position	Company

Phone	Cell	Email

Message	Type of call
	☐ Urgent
	☐ Call
	☐ Returning Call
	☐ Please Call
	☐ Will Call Again
	☐ Came to see you
	☐ Wants to see you

Delivered? ☐	For	Date	Time

Caller	Position	Company

Phone	Cell	Email

Message	Type of call
	☐ Urgent
	☐ Call
	☐ Returning Call
	☐ Please Call
	☐ Will Call Again
	☐ Came to see you
	☐ Wants to see you

Delivered? ☐	For	Date	Time

Caller	Position	Company

Phone	Cell	Email

Message	Type of call
	☐ Urgent
	☐ Call
	☐ Returning Call
	☐ Please Call
	☐ Will Call Again
	☐ Came to see you
	☐ Wants to see you

Delivered? ☐	For	Date	Time

Caller	Position	Company

Phone	Cell	Email

Message	Type of call
	☐ Urgent
	☐ Call
	☐ Returning Call
	☐ Please Call
	☐ Will Call Again
	☐ Came to see you
	☐ Wants to see you

Delivered? ☐	For	Date	Time
Caller	Position		Company
Phone	Cell		Email

Message	Type of call
	☐ Urgent
	☐ Call
	☐ Returning Call
	☐ Please Call
	☐ Will Call Again
	☐ Came to see you
	☐ Wants to see you

Delivered? ☐	For	Date	Time
Caller	Position		Company
Phone	Cell		Email

Message	Type of call
	☐ Urgent
	☐ Call
	☐ Returning Call
	☐ Please Call
	☐ Will Call Again
	☐ Came to see you
	☐ Wants to see you

Delivered? ☐	For	Date	Time
Caller	Position		Company
Phone	Cell		Email

Message	Type of call
	☐ Urgent
	☐ Call
	☐ Returning Call
	☐ Please Call
	☐ Will Call Again
	☐ Came to see you
	☐ Wants to see you

Delivered?	For	Date	Time
☐			

Caller	Position	Company

Phone	Cell	Email

Message	Type of call
	☐ Urgent ☐ Call ☐ Returning Call ☐ Please Call ☐ Will Call Again ☐ Came to see you ☐ Wants to see you

Delivered?	For	Date	Time
☐			

Caller	Position	Company

Phone	Cell	Email

Message	Type of call
	☐ Urgent ☐ Call ☐ Returning Call ☐ Please Call ☐ Will Call Again ☐ Came to see you ☐ Wants to see you

Delivered?	For	Date	Time
☐			

Caller	Position	Company

Phone	Cell	Email

Message	Type of call
	☐ Urgent ☐ Call ☐ Returning Call ☐ Please Call ☐ Will Call Again ☐ Came to see you ☐ Wants to see you

Delivered? ☐	For	Date	Time

Caller	Position	Company

Phone	Cell	Email

Message	Type of call
	☐ Urgent
	☐ Call
	☐ Returning Call
	☐ Please Call
	☐ Will Call Again
	☐ Came to see you
	☐ Wants to see you

Delivered? ☐	For	Date	Time

Caller	Position	Company

Phone	Cell	Email

Message	Type of call
	☐ Urgent
	☐ Call
	☐ Returning Call
	☐ Please Call
	☐ Will Call Again
	☐ Came to see you
	☐ Wants to see you

Delivered? ☐	For	Date	Time

Caller	Position	Company

Phone	Cell	Email

Message	Type of call
	☐ Urgent
	☐ Call
	☐ Returning Call
	☐ Please Call
	☐ Will Call Again
	☐ Came to see you
	☐ Wants to see you

Delivered? ☐	For	Date	Time

Caller	Position	Company

Phone	Cell	Email

Message	Type of call
	☐ Urgent
	☐ Call
	☐ Returning Call
	☐ Please Call
	☐ Will Call Again
	☐ Came to see you
	☐ Wants to see you

Delivered? ☐	For	Date	Time

Caller	Position	Company

Phone	Cell	Email

Message	Type of call
	☐ Urgent
	☐ Call
	☐ Returning Call
	☐ Please Call
	☐ Will Call Again
	☐ Came to see you
	☐ Wants to see you

Delivered? ☐	For	Date	Time

Caller	Position	Company

Phone	Cell	Email

Message	Type of call
	☐ Urgent
	☐ Call
	☐ Returning Call
	☐ Please Call
	☐ Will Call Again
	☐ Came to see you
	☐ Wants to see you

Delivered? ☐	For	Date	Time

Caller	Position	Company

Phone	Cell	Email

Message	Type of call
	☐ Urgent
	☐ Call
	☐ Returning Call
	☐ Please Call
	☐ Will Call Again
	☐ Came to see you
	☐ Wants to see you

Delivered? ☐	For	Date	Time

Caller	Position	Company

Phone	Cell	Email

Message	Type of call
	☐ Urgent
	☐ Call
	☐ Returning Call
	☐ Please Call
	☐ Will Call Again
	☐ Came to see you
	☐ Wants to see you

Delivered? ☐	For	Date	Time

Caller	Position	Company

Phone	Cell	Email

Message	Type of call
	☐ Urgent
	☐ Call
	☐ Returning Call
	☐ Please Call
	☐ Will Call Again
	☐ Came to see you
	☐ Wants to see you

Delivered? ☐	For	Date	Time

Caller	Position	Company

Phone	Cell	Email

Message	Type of call
	☐ Urgent
	☐ Call
	☐ Returning Call
	☐ Please Call
	☐ Will Call Again
	☐ Came to see you
	☐ Wants to see you

Delivered? ☐	For	Date	Time

Caller	Position	Company

Phone	Cell	Email

Message	Type of call
	☐ Urgent
	☐ Call
	☐ Returning Call
	☐ Please Call
	☐ Will Call Again
	☐ Came to see you
	☐ Wants to see you

Delivered? ☐	For	Date	Time

Caller	Position	Company

Phone	Cell	Email

Message	Type of call
	☐ Urgent
	☐ Call
	☐ Returning Call
	☐ Please Call
	☐ Will Call Again
	☐ Came to see you
	☐ Wants to see you

Delivered? ☐	For	Date	Time

Caller	Position	Company

Phone	Cell	Email

Message	Type of call
	☐ Urgent
	☐ Call
	☐ Returning Call
	☐ Please Call
	☐ Will Call Again
	☐ Came to see you
	☐ Wants to see you

Delivered? ☐	For	Date	Time

Caller	Position	Company

Phone	Cell	Email

Message	Type of call
	☐ Urgent
	☐ Call
	☐ Returning Call
	☐ Please Call
	☐ Will Call Again
	☐ Came to see you
	☐ Wants to see you

Delivered? ☐	For	Date	Time

Caller	Position	Company

Phone	Cell	Email

Message	Type of call
	☐ Urgent
	☐ Call
	☐ Returning Call
	☐ Please Call
	☐ Will Call Again
	☐ Came to see you
	☐ Wants to see you

Delivered? ☐	For	Date	Time

Caller	Position	Company

Phone	Cell	Email

Message	Type of call
	☐ Urgent
	☐ Call
	☐ Returning Call
	☐ Please Call
	☐ Will Call Again
	☐ Came to see you
	☐ Wants to see you

Delivered? ☐	For	Date	Time

Caller	Position	Company

Phone	Cell	Email

Message	Type of call
	☐ Urgent
	☐ Call
	☐ Returning Call
	☐ Please Call
	☐ Will Call Again
	☐ Came to see you
	☐ Wants to see you

Delivered? ☐	For	Date	Time

Caller	Position	Company

Phone	Cell	Email

Message	Type of call
	☐ Urgent
	☐ Call
	☐ Returning Call
	☐ Please Call
	☐ Will Call Again
	☐ Came to see you
	☐ Wants to see you

Delivered?	For	Date	Time
☐			

Caller	Position	Company

Phone	Cell	Email

Message	Type of call
	☐ Urgent ☐ Call ☐ Returning Call ☐ Please Call ☐ Will Call Again ☐ Came to see you ☐ Wants to see you

Delivered?	For	Date	Time
☐			

Caller	Position	Company

Phone	Cell	Email

Message	Type of call
	☐ Urgent ☐ Call ☐ Returning Call ☐ Please Call ☐ Will Call Again ☐ Came to see you ☐ Wants to see you

Delivered?	For	Date	Time
☐			

Caller	Position	Company

Phone	Cell	Email

Message	Type of call
	☐ Urgent ☐ Call ☐ Returning Call ☐ Please Call ☐ Will Call Again ☐ Came to see you ☐ Wants to see you

Delivered? ☐	For	Date	Time

Caller	Position	Company

Phone	Cell	Email

Message	Type of call
	☐ Urgent ☐ Call ☐ Returning Call ☐ Please Call ☐ Will Call Again ☐ Came to see you ☐ Wants to see you

Delivered? ☐	For	Date	Time

Caller	Position	Company

Phone	Cell	Email

Message	Type of call
	☐ Urgent ☐ Call ☐ Returning Call ☐ Please Call ☐ Will Call Again ☐ Came to see you ☐ Wants to see you

Delivered? ☐	For	Date	Time

Caller	Position	Company

Phone	Cell	Email

Message	Type of call
	☐ Urgent ☐ Call ☐ Returning Call ☐ Please Call ☐ Will Call Again ☐ Came to see you ☐ Wants to see you

Delivered? ☐	For	Date	Time
Caller	Position		Company
Phone	Cell		Email

Message	Type of call
	☐ Urgent ☐ Call ☐ Returning Call ☐ Please Call ☐ Will Call Again ☐ Came to see you ☐ Wants to see you

Delivered? ☐	For	Date	Time
Caller	Position		Company
Phone	Cell		Email

Message	Type of call
	☐ Urgent ☐ Call ☐ Returning Call ☐ Please Call ☐ Will Call Again ☐ Came to see you ☐ Wants to see you

Delivered? ☐	For	Date	Time
Caller	Position		Company
Phone	Cell		Email

Message	Type of call
	☐ Urgent ☐ Call ☐ Returning Call ☐ Please Call ☐ Will Call Again ☐ Came to see you ☐ Wants to see you

Delivered? ☐	For	Date	Time
Caller	Position		Company
Phone	Cell		Email

Message	Type of call
	☐ Urgent
	☐ Call
	☐ Returning Call
	☐ Please Call
	☐ Will Call Again
	☐ Came to see you
	☐ Wants to see you

Delivered? ☐	For	Date	Time
Caller	Position		Company
Phone	Cell		Email

Message	Type of call
	☐ Urgent
	☐ Call
	☐ Returning Call
	☐ Please Call
	☐ Will Call Again
	☐ Came to see you
	☐ Wants to see you

Delivered? ☐	For	Date	Time
Caller	Position		Company
Phone	Cell		Email

Message	Type of call
	☐ Urgent
	☐ Call
	☐ Returning Call
	☐ Please Call
	☐ Will Call Again
	☐ Came to see you
	☐ Wants to see you

Delivered? ☐	For	Date	Time

Caller	Position	Company

Phone	Cell	Email

Message	Type of call
	☐ Urgent
	☐ Call
	☐ Returning Call
	☐ Please Call
	☐ Will Call Again
	☐ Came to see you
	☐ Wants to see you

Delivered? ☐	For	Date	Time

Caller	Position	Company

Phone	Cell	Email

Message	Type of call
	☐ Urgent
	☐ Call
	☐ Returning Call
	☐ Please Call
	☐ Will Call Again
	☐ Came to see you
	☐ Wants to see you

Delivered? ☐	For	Date	Time

Caller	Position	Company

Phone	Cell	Email

Message	Type of call
	☐ Urgent
	☐ Call
	☐ Returning Call
	☐ Please Call
	☐ Will Call Again
	☐ Came to see you
	☐ Wants to see you

Delivered? ☐	For	Date	Time

Caller	Position	Company

Phone	Cell	Email

Message	Type of call
	☐ Urgent ☐ Call ☐ Returning Call ☐ Please Call ☐ Will Call Again ☐ Came to see you ☐ Wants to see you

Delivered? ☐	For	Date	Time

Caller	Position	Company

Phone	Cell	Email

Message	Type of call
	☐ Urgent ☐ Call ☐ Returning Call ☐ Please Call ☐ Will Call Again ☐ Came to see you ☐ Wants to see you

Delivered? ☐	For	Date	Time

Caller	Position	Company

Phone	Cell	Email

Message	Type of call
	☐ Urgent ☐ Call ☐ Returning Call ☐ Please Call ☐ Will Call Again ☐ Came to see you ☐ Wants to see you

Delivered? ☐	For	Date	Time

Caller	Position	Company

Phone	Cell	Email

Message	Type of call
	☐ Urgent
	☐ Call
	☐ Returning Call
	☐ Please Call
	☐ Will Call Again
	☐ Came to see you
	☐ Wants to see you

Delivered? ☐	For	Date	Time

Caller	Position	Company

Phone	Cell	Email

Message	Type of call
	☐ Urgent
	☐ Call
	☐ Returning Call
	☐ Please Call
	☐ Will Call Again
	☐ Came to see you
	☐ Wants to see you

Delivered? ☐	For	Date	Time

Caller	Position	Company

Phone	Cell	Email

Message	Type of call
	☐ Urgent
	☐ Call
	☐ Returning Call
	☐ Please Call
	☐ Will Call Again
	☐ Came to see you
	☐ Wants to see you

Delivered? ☐	For	Date	Time

Caller	Position	Company

Phone	Cell	Email

Message	Type of call
	☐ Urgent ☐ Call ☐ Returning Call ☐ Please Call ☐ Will Call Again ☐ Came to see you ☐ Wants to see you

Delivered? ☐	For	Date	Time

Caller	Position	Company

Phone	Cell	Email

Message	Type of call
	☐ Urgent ☐ Call ☐ Returning Call ☐ Please Call ☐ Will Call Again ☐ Came to see you ☐ Wants to see you

Delivered? ☐	For	Date	Time

Caller	Position	Company

Phone	Cell	Email

Message	Type of call
	☐ Urgent ☐ Call ☐ Returning Call ☐ Please Call ☐ Will Call Again ☐ Came to see you ☐ Wants to see you

Delivered? ☐	For	Date	Time

Caller	Position	Company

Phone	Cell	Email

Message	Type of call
	☐ Urgent
	☐ Call
	☐ Returning Call
	☐ Please Call
	☐ Will Call Again
	☐ Came to see you
	☐ Wants to see you

Delivered? ☐	For	Date	Time

Caller	Position	Company

Phone	Cell	Email

Message	Type of call
	☐ Urgent
	☐ Call
	☐ Returning Call
	☐ Please Call
	☐ Will Call Again
	☐ Came to see you
	☐ Wants to see you

Delivered? ☐	For	Date	Time

Caller	Position	Company

Phone	Cell	Email

Message	Type of call
	☐ Urgent
	☐ Call
	☐ Returning Call
	☐ Please Call
	☐ Will Call Again
	☐ Came to see you
	☐ Wants to see you

Delivered? ☐	For	Date	Time

Caller	Position	Company

Phone	Cell	Email

Message	Type of call
	☐ Urgent
	☐ Call
	☐ Returning Call
	☐ Please Call
	☐ Will Call Again
	☐ Came to see you
	☐ Wants to see you

Delivered? ☐	For	Date	Time

Caller	Position	Company

Phone	Cell	Email

Message	Type of call
	☐ Urgent
	☐ Call
	☐ Returning Call
	☐ Please Call
	☐ Will Call Again
	☐ Came to see you
	☐ Wants to see you

Delivered? ☐	For	Date	Time

Caller	Position	Company

Phone	Cell	Email

Message	Type of call
	☐ Urgent
	☐ Call
	☐ Returning Call
	☐ Please Call
	☐ Will Call Again
	☐ Came to see you
	☐ Wants to see you

Delivered? ☐	For	Date	Time

Caller	Position	Company

Phone	Cell	Email

Message	Type of call
	☐ Urgent
	☐ Call
	☐ Returning Call
	☐ Please Call
	☐ Will Call Again
	☐ Came to see you
	☐ Wants to see you

Delivered? ☐	For	Date	Time

Caller	Position	Company

Phone	Cell	Email

Message	Type of call
	☐ Urgent
	☐ Call
	☐ Returning Call
	☐ Please Call
	☐ Will Call Again
	☐ Came to see you
	☐ Wants to see you

Delivered? ☐	For	Date	Time

Caller	Position	Company

Phone	Cell	Email

Message	Type of call
	☐ Urgent
	☐ Call
	☐ Returning Call
	☐ Please Call
	☐ Will Call Again
	☐ Came to see you
	☐ Wants to see you

Delivered? ☐	For	Date	Time

Caller	Position	Company

Phone	Cell	Email

Message	Type of call
	☐ Urgent
	☐ Call
	☐ Returning Call
	☐ Please Call
	☐ Will Call Again
	☐ Came to see you
	☐ Wants to see you

Delivered? ☐	For	Date	Time

Caller	Position	Company

Phone	Cell	Email

Message	Type of call
	☐ Urgent
	☐ Call
	☐ Returning Call
	☐ Please Call
	☐ Will Call Again
	☐ Came to see you
	☐ Wants to see you

Delivered? ☐	For	Date	Time

Caller	Position	Company

Phone	Cell	Email

Message	Type of call
	☐ Urgent
	☐ Call
	☐ Returning Call
	☐ Please Call
	☐ Will Call Again
	☐ Came to see you
	☐ Wants to see you

Delivered? ☐	For	Date	Time

Caller	Position	Company

Phone	Cell	Email

Message	Type of call
	☐ Urgent
	☐ Call
	☐ Returning Call
	☐ Please Call
	☐ Will Call Again
	☐ Came to see you
	☐ Wants to see you

Delivered? ☐	For	Date	Time

Caller	Position	Company

Phone	Cell	Email

Message	Type of call
	☐ Urgent
	☐ Call
	☐ Returning Call
	☐ Please Call
	☐ Will Call Again
	☐ Came to see you
	☐ Wants to see you

Delivered? ☐	For	Date	Time

Caller	Position	Company

Phone	Cell	Email

Message	Type of call
	☐ Urgent
	☐ Call
	☐ Returning Call
	☐ Please Call
	☐ Will Call Again
	☐ Came to see you
	☐ Wants to see you

Delivered? ☐	For	Date	Time

Caller	Position	Company

Phone	Cell	Email

Message	Type of call
	☐ Urgent
	☐ Call
	☐ Returning Call
	☐ Please Call
	☐ Will Call Again
	☐ Came to see you
	☐ Wants to see you

Delivered? ☐	For	Date	Time

Caller	Position	Company

Phone	Cell	Email

Message	Type of call
	☐ Urgent
	☐ Call
	☐ Returning Call
	☐ Please Call
	☐ Will Call Again
	☐ Came to see you
	☐ Wants to see you

Delivered? ☐	For	Date	Time

Caller	Position	Company

Phone	Cell	Email

Message	Type of call
	☐ Urgent
	☐ Call
	☐ Returning Call
	☐ Please Call
	☐ Will Call Again
	☐ Came to see you
	☐ Wants to see you

Delivered? ☐	For	Date	Time

Caller	Position	Company

Phone	Cell	Email

Message	Type of call
	☐ Urgent
	☐ Call
	☐ Returning Call
	☐ Please Call
	☐ Will Call Again
	☐ Came to see you
	☐ Wants to see you

Delivered? ☐	For	Date	Time

Caller	Position	Company

Phone	Cell	Email

Message	Type of call
	☐ Urgent
	☐ Call
	☐ Returning Call
	☐ Please Call
	☐ Will Call Again
	☐ Came to see you
	☐ Wants to see you

Delivered? ☐	For	Date	Time

Caller	Position	Company

Phone	Cell	Email

Message	Type of call
	☐ Urgent
	☐ Call
	☐ Returning Call
	☐ Please Call
	☐ Will Call Again
	☐ Came to see you
	☐ Wants to see you

Delivered? ☐	For	Date	Time

Caller	Position	Company

Phone	Cell	Email

Message	Type of call
	☐ Urgent
	☐ Call
	☐ Returning Call
	☐ Please Call
	☐ Will Call Again
	☐ Came to see you
	☐ Wants to see you

Delivered? ☐	For	Date	Time

Caller	Position	Company

Phone	Cell	Email

Message	Type of call
	☐ Urgent
	☐ Call
	☐ Returning Call
	☐ Please Call
	☐ Will Call Again
	☐ Came to see you
	☐ Wants to see you

Delivered? ☐	For	Date	Time

Caller	Position	Company

Phone	Cell	Email

Message	Type of call
	☐ Urgent
	☐ Call
	☐ Returning Call
	☐ Please Call
	☐ Will Call Again
	☐ Came to see you
	☐ Wants to see you